Drawing Lines

1. <u>Draw</u> many <u>lines</u>.

2. Draw some <u>long</u> lines.

3. Put your <u>straightedge</u> along this line. Draw this line as long as you can.

Why can't you make it longer? _ _ _ _ _ _ _ _ _ _ _ _ _ _ _ _ _

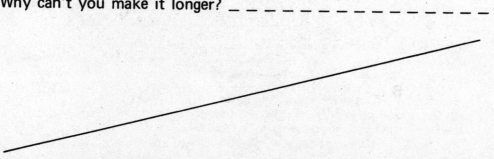

Drawing Lines Through Points

1. Draw a line through the point.

.

2. Draw three lines through point A.

. A

3. Draw a line through points B and C. Make it as long as you can.

. C

B .

1. Draw many lines through the point.

•

2. How many lines can you draw through a point? _ _ _ _ _ _ _ _ _ _ _

_ _

3. Draw a long line through the two points, D and E.

•
E

•
D

Can you draw a __different__ line through the two points? _ _ _ _ _

4. Do these three points __lie on__ the __same__ line? _ _ _ _ _

•
O

•
P

•
Q

Line Segments

1. Draw a line segment from point A to point B.

A •

•

• B

2. Connect P and Q by a line segment.

• Q

P •

3. Can you draw another line segment between P and Q which is different

from the first one you drew? _ _ _ _ _

4. Draw some line segments.

1. <u>Label</u> as R and S the <u>endpoints</u> of the given segment.

2. Draw a long line through R and S. How are line segment $\overline{RS}$ and line $\overleftrightarrow{RS}$ different? _

_ _

3. Draw a line through R which does not pass through S.

4. Connect A and B by a line segment.

A .

• B

5. Draw a line through F and G.

•
F

•
G

6. How is a line segment different from a line? _ _ _ _ _ _ _ _ _ _ _

_ _

_ _

<u>Triangles</u>

1. Draw four lines through point A.

• **A**

2. Connect X and Y by a line segment.

• **Y**

X •

3. Draw segment $\overline{AB}$.
   Draw segment $\overline{AC}$.
   Draw segment $\overline{BC}$.

• **B**

• **C**

A •

This <u>figure</u> with three <u>sides</u> is a <u>triangle</u>. Each side is a line segment.

1. Connect P and Q by a line segment.

P •

•
Q

2. Connect A and B.

   Connect B and C.

   Connect A and C.

•
C

A •

• B

3. What kind of figure is ABC? _ _ _ _ _ _ _ _ _ _ _ _ _ _ _ _ _

   (a) line   (c) segment

   (b) triangle   (d) four

1. Name the segment. _ _ _ _ _ _

2. Name it a different way. _ _ _ _ _ _

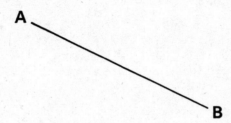

3. Name the triangle. _ _ _ _ _ _

4. Name it some different ways. _ _ _ _ _ _ _ _ _ _ _ _ _ _ _ _

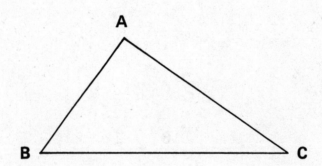

<u>Quadrilaterals</u>

1. Connect Q and R.
2. Connect Q and P.
3. Connect P and S.
4. Connect R and S.

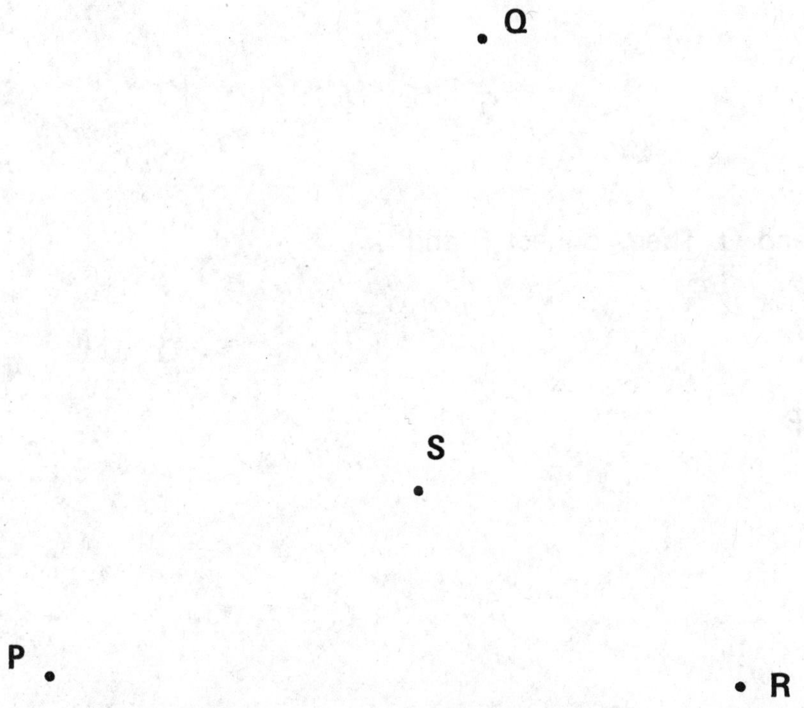

. Q

S
.

P .

. R

5. How many sides does the figure have? _ _ _ _ _

1. Draw the triangle ABC by drawing line segments $\overline{AB}$, $\overline{BC}$, $\overline{CA}$.

. B

. C

A .

2. Connect P and Q. Then connect R and S.

. Q

P .

R .

. S

3. Now connect P and R. Then connect Q and S.

This figure with four sides is a <u>quadrilateral</u>. Each side is a line segment.

4. Draw another quadrilateral.

Pentagons

1. Given the points A, B, C, D, and E, connect A and B.

A.

E.   . B

D.   . C

2. Draw $\overline{BC}$. Then draw $\overline{CD}$. Then draw $\overline{DE}$. Then draw $\overline{EA}$.

3. How many sides has figure ABCDE? _ _ _ _ _
   This figure with five sides is a <u>pentagon</u>.

4. Draw another pentagon.

5. Label the <u>corners</u> of the pentagon with letters.

1. Draw triangle ABC.

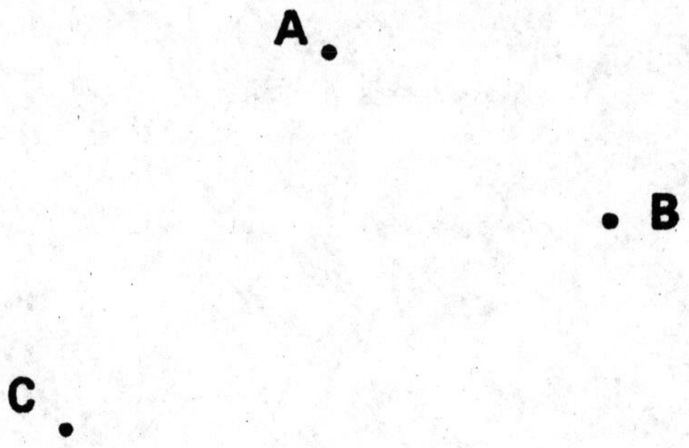

A .

. B

C .

2. Draw quadrilateral QPSR.

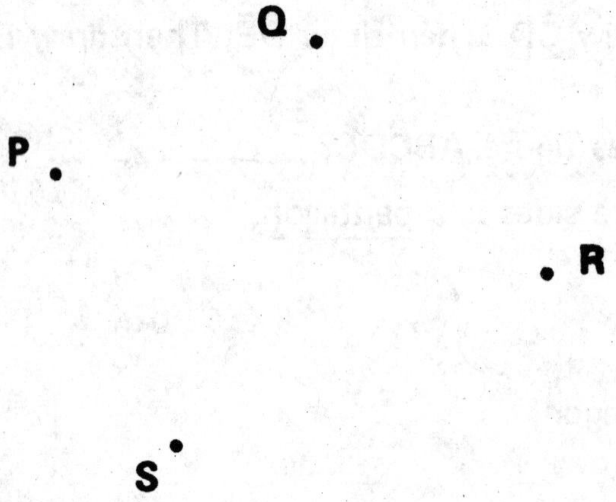

Q .

P .

. R

S .

3. Draw pentagon UWXYZ.

U .

. W

Z .

. X

. Y

Connecting Points

1. <u>Given</u> points A, B, and C, draw all the line segments you can which connect any two of them.

.B

A .

.
C

2. How many sides does a triangle have? _ _ _ _ _

3. Given points P, Q, R, and S, draw all the line segments which connect any two of them.

.Q

P .

. R

S .

You drew _ _ _ _ _ _ _ _ _ _ _ _ _ _ _ segments.
  (a) three   (c) five
  (b) four   (d) six

4. How many segments did you draw? _ _ _ _ _

5. Name the segments you drew. _

1. Given points A, B, C, and D, draw all line segments which connect any two of them.

B

A

C

D

2. How many segments did you draw? _ _ _ _ _

3. Given points E, F, and G, draw all the line segments which connect two of them.

F

E

G

4. How many segments did you draw? _ _ _ _ _

5. Name the segments you drew. _ _ _ _ _ _ _ _ _ _ _ _ _ _ _

Review

1. Draw three points on your paper. Connect them so that they make a triangle.

2. Draw four points L, M, N, O on your paper so that they will <u>form</u> a quadrilateral when you connect them.

3. Draw five points P, Q, R, S, T on your paper so that they will form a pentagon when you connect them.

1. Match each word with the figure it names.

   point

   line segment

   triangle

   quadrilateral

2. A line segment _ _ _ _ _ _ _ _ _ _ _ _ _ _ _ _ _ _ two points.

   (a) is   (c) draws

   (b) connects   (d) labels

3. Draw triangle BDE.

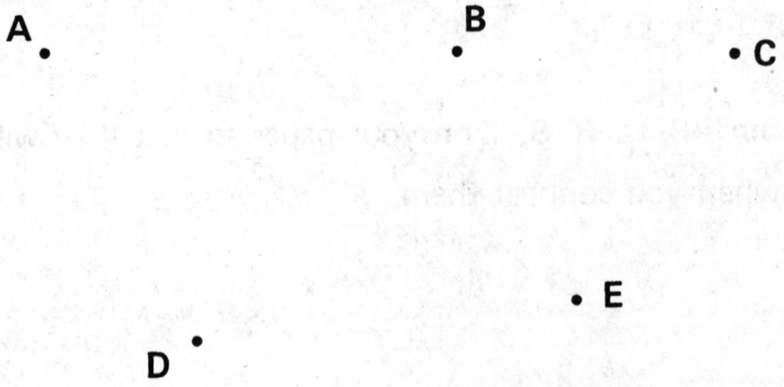

   A.

                    B

                             • C

                                • E

   D

4. Draw line segment $\overline{AC}$ in the figure above.

Points of Intersection

1. Draw segment $\overline{AB}$.
   Draw segment $\overline{CD}$.

• D

A •

C •

• B

2. The point of intersection is the point where the two segments meet.
   Label as E the point of intersection of $\overline{AB}$ and $\overline{CD}$.

3. Draw another line segment through E.

4. Draw two lines.
   Make them intersect.

Label all the points of intersection with letters.

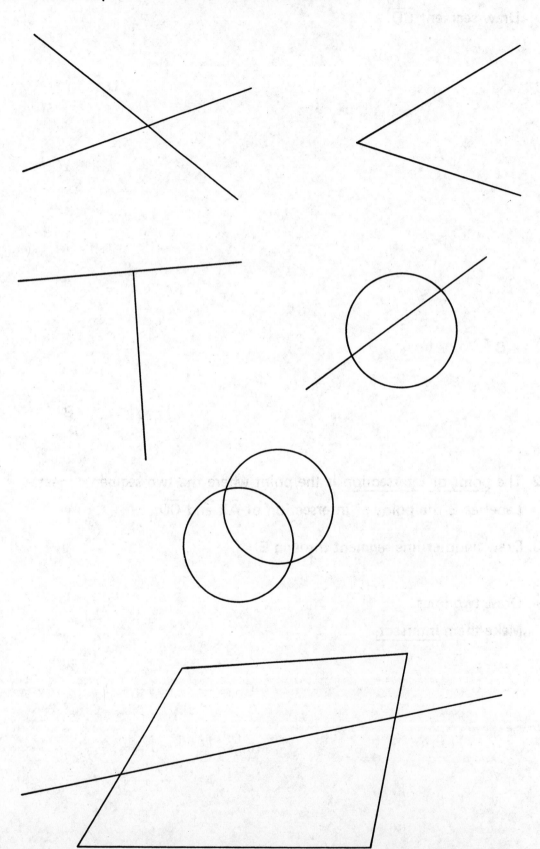

1. Draw segment $\overline{AB}$.
   Draw segment $\overline{CD}$.
   Draw segment $\overline{EF}$.

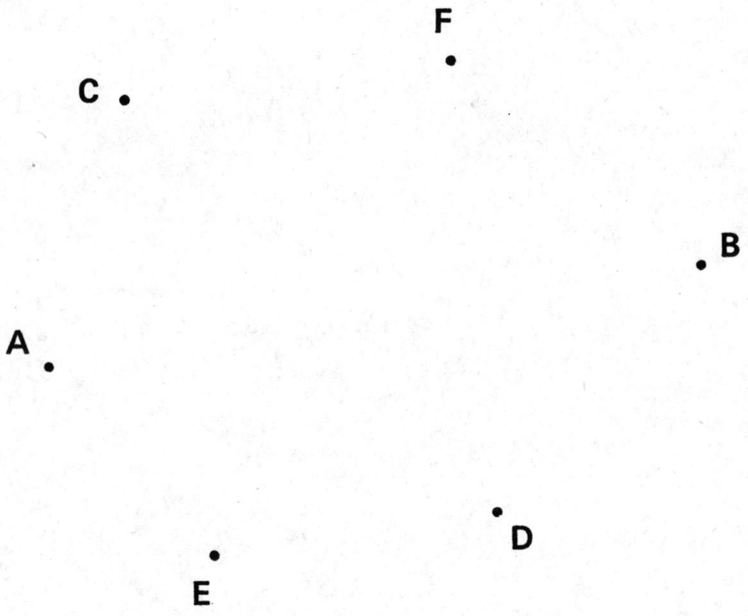

F
•

C •

•B

A •

•D

•
E

2. How many points of intersection are there? _ _ _ _ _

3. Draw two line segments which do not intersect.

1. Given A, B, C, D, and E, draw all line segments you can which connect two of them.

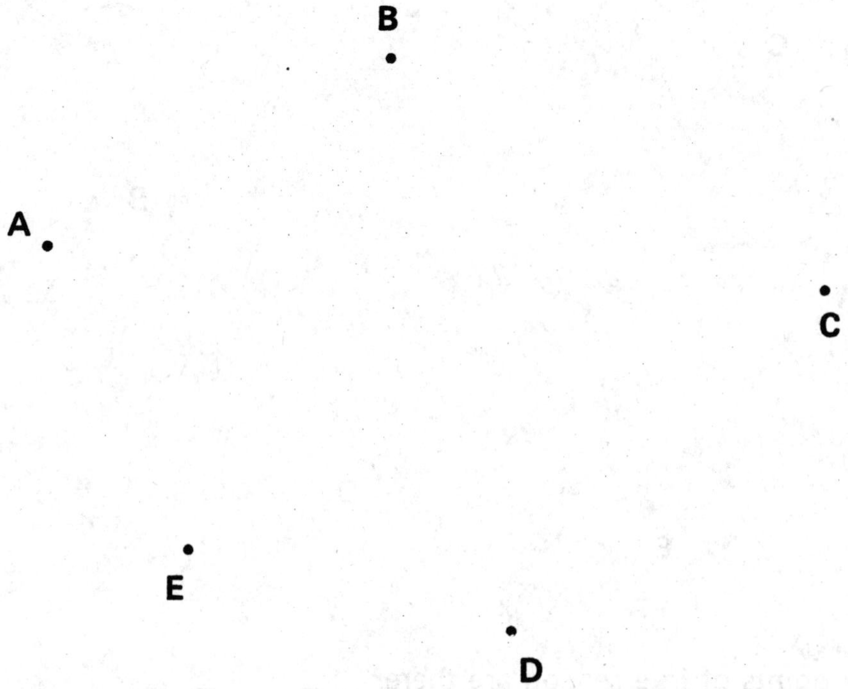

B
•

A
•

•
C

•
E

•
D

2. How many line segments did you draw? _ _ _ _ _

3. How many points of intersection are there? _ _ _ _ _

The Line Through Two Points

1. Draw a line through A and B.

• B

• A

2. Draw the line segment $\overline{XY}$.

X •

• Y

Ż

3. Draw the line through X and Z.

4. Can you draw a different line through X and Z? _ _ _ _ _ _

1. Draw the triangle ABC.

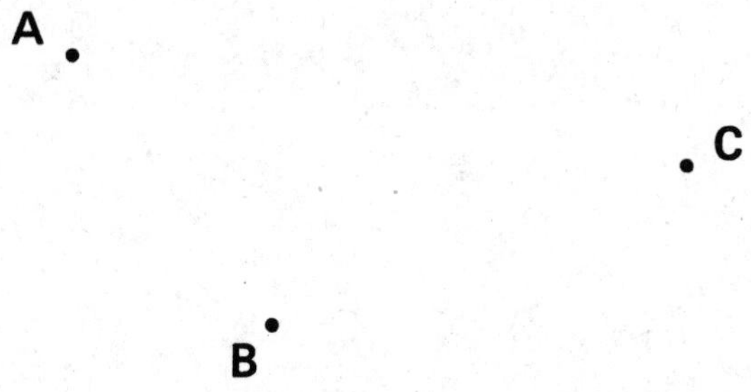

A .

. C

.
B

2. Draw 4 lines through the point X.

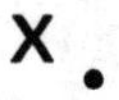

X .

3. Label as P and Q the endpoints of the segment.

4. Given points U, V, W, and Z, draw all the segments which connect two of them.

U .

. W

V .

. Z

1. Draw $\overline{AB}$. Then draw $\overline{AC}$. Then draw $\overline{CD}$.

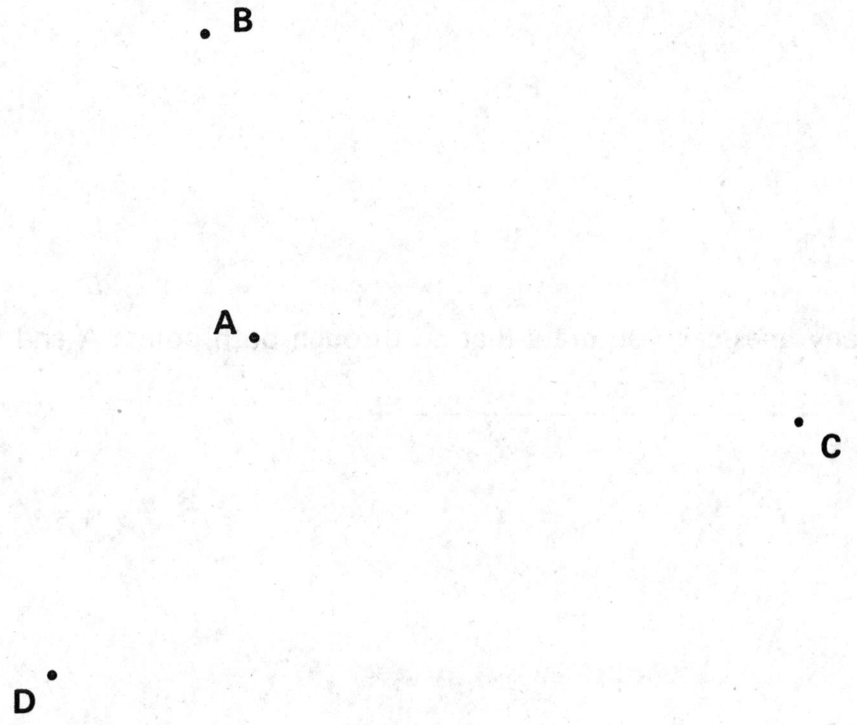

. B

A .

. C

. D

2. Do these three line segments form a triangle? _ _ _ _ _

3. Now draw $\overline{AD}$.

4. Do these four line segments form a quadrilateral? _ _ _ _ _

5. Draw a quadrilateral.

The Line Determined by Two Points

1. How many lines can you draw through point P? _ _ _ _ _ _ _ _
_ _

P .

2. How many lines can you draw that go through both points A and B?
_ _ _ _ _ _ _ _ _ _ _ _ _ _ _

. B

. A

3. Draw the line determined by points S and T.

. S

. T

4. Draw the line segment determined by X and Y.

. Y

. X

1. Draw the line determined by P and Q.

P .

. Q

2. Draw all the lines determined by <u>pairs</u> of the points X, Y, Z, and W.

. Y

X .

. Z

W .

3. Draw the triangle determined by A, B, and C.

. B

A .

C

1. How many points <u>determine</u> a line? _ _ _ _ _ _ _

2. Given points A, B, C, and D, draw all the segments which connect pairs of them.

. B

A .

. D

. C

3. How many points of intersection are there in problem 2? _ _ _ _ _ _

4. Draw all the line segments determined by pairs of the points P, Q, R, X, and Y.

P .   . Q

. R

X .

. Y

5. How many points of intersection are there in problem 4? _ _ _ _ _ _

C

B

A

D

E

F

1. Are points A, B, and C on the same line? _ _ _ _ _

   Are points D, E, and F on the same line? _ _ _ _ _

2. Draw lines $\overleftrightarrow{AE}$ and $\overleftrightarrow{BD}$.

   Label their point of intersection X.

3. Draw lines $\overleftrightarrow{AF}$ and $\overleftrightarrow{CD}$.

   Label their point of intersection Y.

4. Draw lines $\overleftrightarrow{BF}$ and $\overleftrightarrow{CE}$.

   Label their point of intersection Z.

5. Are points X, Y, and Z on the same line? _ _ _ _ _

Triangles

1. Draw the line segment determined by A and B.

Draw the line segment determined by C and D.

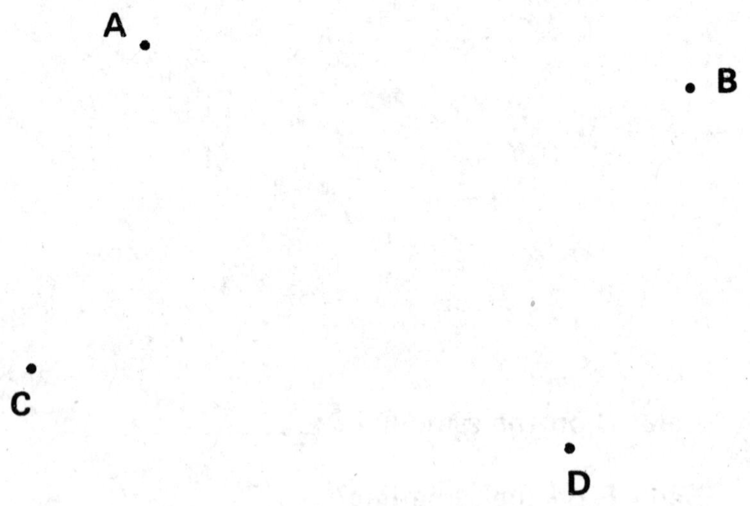

A .

. B

.
C

.
D

2. Do these line segments $\overline{AB}$ and $\overline{CD}$ intersect? _ _ _ _ _

3. Draw $\overline{AD}$ and $\overline{BC}$.

4. How many triangles do you see? _ _ _ _ _

Triangles

1. Draw $\overline{AV}$.

.V

A .

.
C

.
W

2. Draw $\overline{CW}$.

Draw $\overline{AC}$ and $\overline{WC}$.

Draw $\overline{VW}$ and $\overline{CA}$.

3. Connect C and V.

4. Name the triangles you see. _ _ _ _ _ _ _ _ _ _ _ _ _ _ _ _ _

Draw pictures to help you answer these questions.

1. How many line segments can you draw which connect two points?

_ _ _ _ _

2. Given three points, how many line segments can you draw which connect two of them? _ _ _ _ _ _

3. Given four points, how many line segments can you draw which connect two of them? _ _ _ _ _ _

4. Given five points, how many line segments can you draw which connect two of them? _ _ _ _ _ _

Vocabulary Quiz

1. Match each word with the figure it names.

   line segment

   triangle

   quadrilateral

   point of intersection

   circle

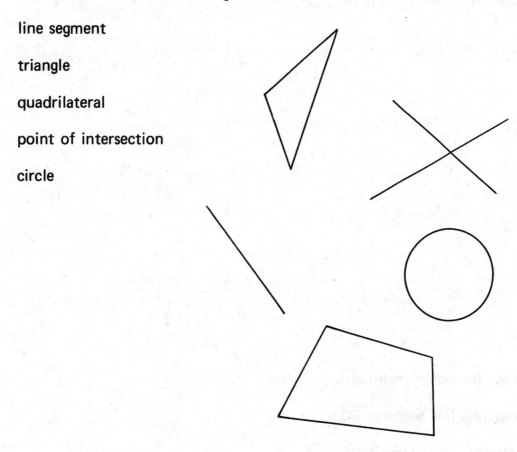

2. The line segment _ _ _ _ _ _ _ _ _ _ _ _ _ _ _ _ points P and Q.
   (a) likes
   (b) moves
   (c) connects
   (d) labels
   (e) draws

3. Two points _ _ _ _ _ _ _ _ _ _ _ _ _ _ a line.

   (a) hold   (d) intersect
   (b) bend   (e) determine
   (c) draw   (f) quadrilateral

Naming Triangles

1. Draw the line segment $\overline{AD}$.

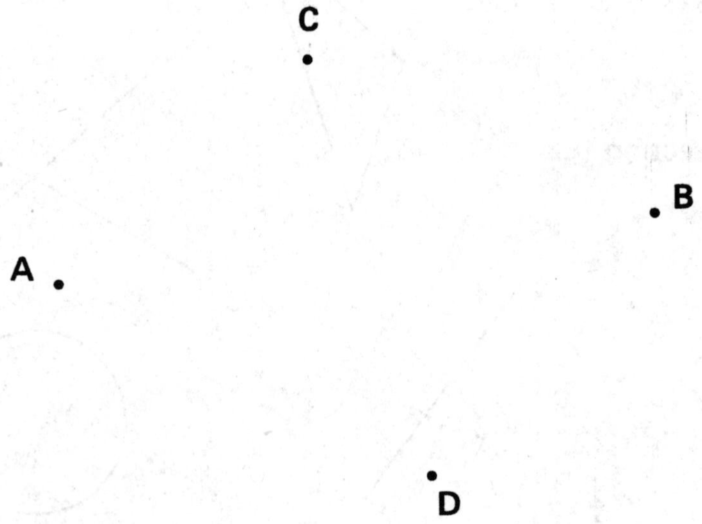

C
•

•B

A •

•
D

2. Draw the line segment $\overline{BD}$.

3. Draw the line segment $\overline{BC}$.

4. Draw the line segment $\overline{CA}$.

5. Draw the line segment $\overline{AB}$.

6. How many triangles do you see in the figure? _ _ _ _ _

7. Name the triangles you see. _ _ _ _ _ _ _ _ _ _ _ _ _ _ _ _

1. Draw the line segment determined by A and D.

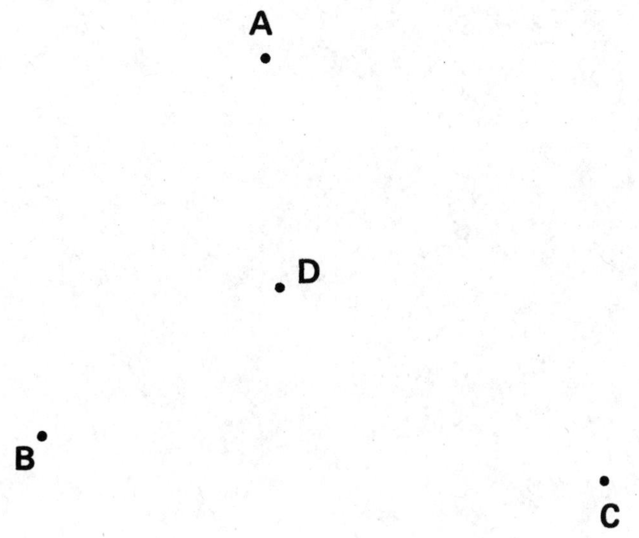

A
•

•D

B•

•
C

2. Draw the line segment determined by B and D.

3. Draw the line segment determined by C and D.

4. Draw the line segment determined by B and C.

5. Draw the line segment determined by A and B.

6. Draw the line segment determined by A and C.

7. How many triangles do you see in the figure? _ _ _ _ _

Name the triangles. _

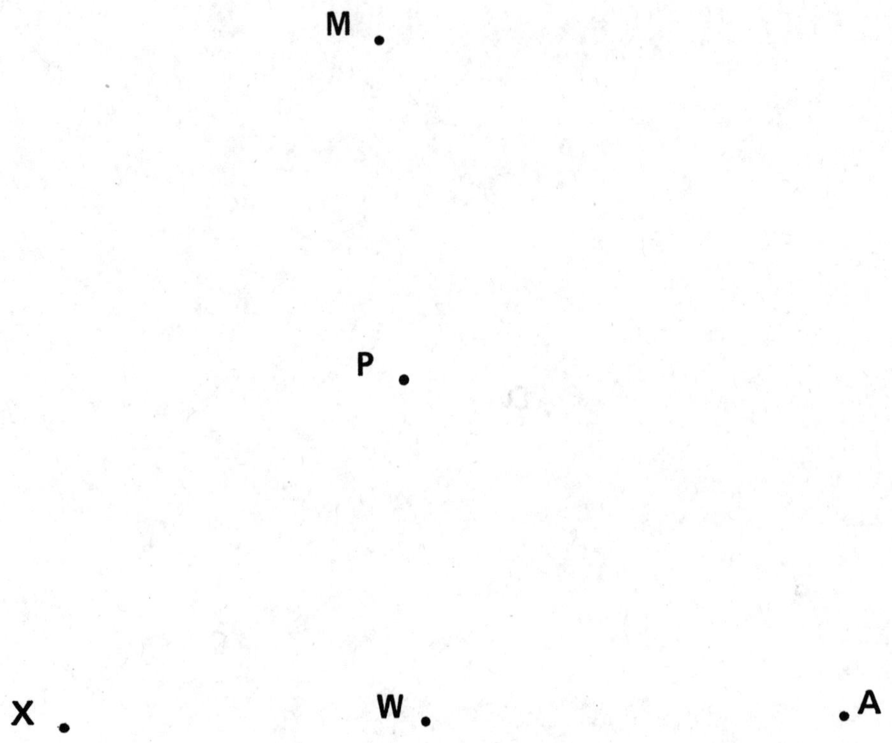

M .

P .

X .   W .   . A

1. Draw $\overline{XP}$. Then draw $\overline{AP}$. Then draw $\overline{MW}$. Then draw $\overline{XA}$.
   Then draw $\overline{MX}$. Then draw $\overline{MA}$.

2. Does $\overline{MW}$ pass through P? _ _ _ _ _

3. Does $\overline{XA}$ pass through W? _ _ _ _ _

4. How many triangles can you find in the figure? _ _ _ _ _

5. Name the triangles you can find. _ _ _ _ _ _ _ _ _ _ _ _ _ _ _ _ _

   _

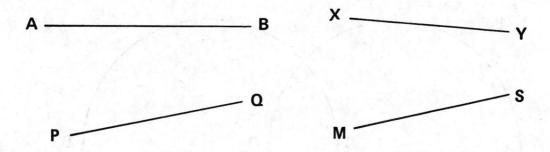

1. Connect A and P. Then connect B and Q.

2. Connect X and M. Then connect Y and S.

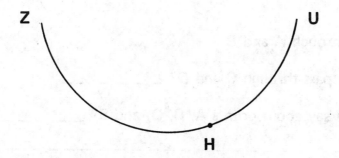

3. Draw the line segment determined by Z and H.

4. Draw $\overline{HU}$. Then draw $\overline{ZU}$.

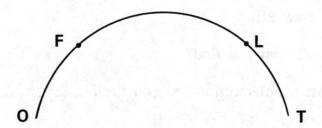

5. Draw $\overline{FL}$. Then draw $\overline{LO}$. Then draw $\overline{OT}$. Then draw $\overline{TF}$.

6. How many triangles do you see in this figure? _ _ _ _ _

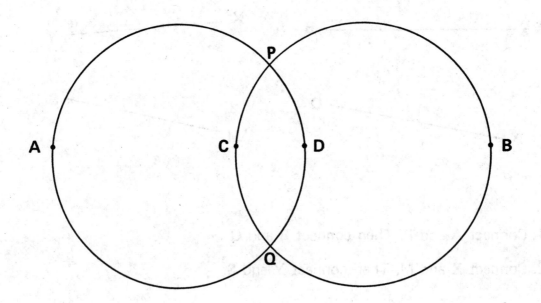

1. Draw a line through A and B.

2. Does this line pass through C and D? _ _ _ _ _ _

3. What can you say about points A, C, D, and B? _ _ _ _ _ _ _ _ _ _

_ _

4. Draw $\overline{PD}$ and $\overline{PB}$.

5. Name the triangle in the figure. _ _ _ _ _ _

6. Draw $\overline{PC}$. Then draw $\overline{PQ}$.

7. How many triangles can you find? _ _ _ _ _ _

8. How many points of intersection can you find? _ _ _ _ _ _

Pentagons

1. Draw a triangle.

2. Draw a quadrilateral.

How many sides does a quadrilateral have? _ _ _ _ _

3. Connect J and L. Then draw $\overline{LN}$. Then draw $\overline{NP}$. Then connect P and R.

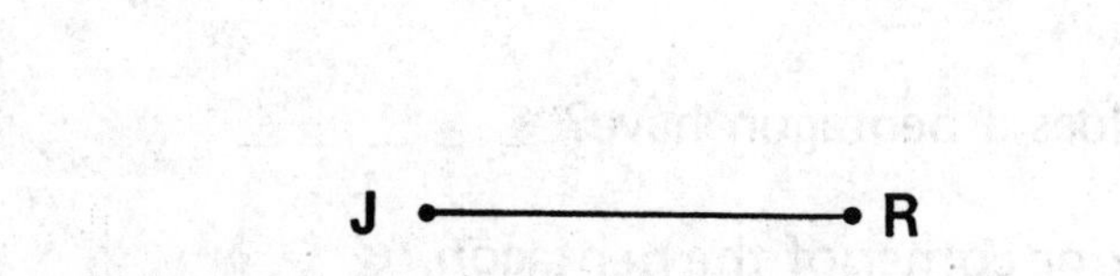

•N

L •

•P

J •————————• R

This figure is a pentagon.

4. How many sides does a pentagon have? _ _ _ _ _

1. Draw the triangle determined by points A, B, and C.

• B

A •

• C

2. Draw the quadrilateral RSTU.

S •

• T

•
R

•
U

3. Draw the pentagon LMNOP.

• M

L •

• N

•
O

•
P

4. How many sides does a pentagon have? _ _ _ _ _

5. Point L is a <u>vertex</u> or corner of the pentagon.

How many <u>vertices</u> does a pentagon have? _ _ _ _ _

Hexagons

1. Connect A and B. Then connect B and C.
   Then connect C and D. Then connect D and E.
   Then connect E and F. Then connect F and A.

• C

B •   • D

• A

• E

• F

2. This figure is a <u>hexagon</u>.

   How many sides does it have? _ _ _ _ _

   How many vertices or corners does it have? _ _ _ _ _

3. Draw hexagon QRSTUV.

R
•

Q •   • S

V •   • T

• U

1. Draw a pentagon. It will have _ _ _ _ _ _ _ _ sides and

_ _ _ _ _ _ _ _ vertices or corners.

2. Draw a hexagon. It will have _ _ _ _ _ _ sides and _ _ _ _ _ _ vertices.

3. Draw a figure with more than six sides.

1. Draw all segments determined by two of the points W, X, Y, and Z.

Y .   . Z

X .   . W

2. How many line segments can you draw which connect four points?

_ _ _ _ _

3. Draw all line segments determined by two of the points M, N, and P.

M .

N .   . P

4. How many line segments can you draw which connect three points?

_ _ _ _ _

5. What do we call this figure? _ _ _ _ _ _ _ _ _ _ _ _ _

6. How many points determine a triangle? _ _ _ _ _

How many points determine a line? _ _ _ _ _

1. Draw all lines determined by two of the points A, B, C, and D.

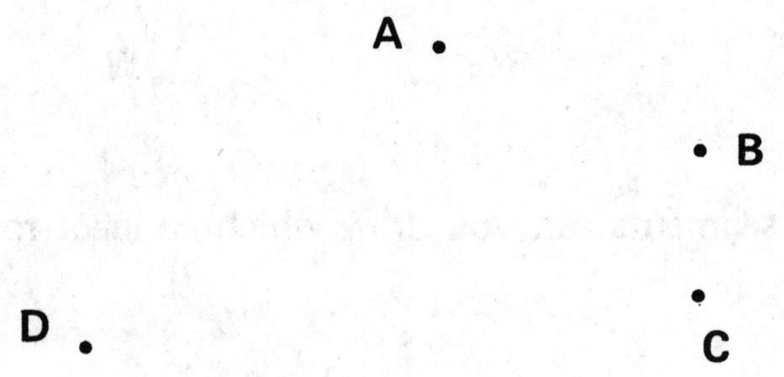

A .

• B

•

D .

•

C

2. How many lines can you draw which connect four points? _ _ _ _ _

3. How many points of intersection are there? _ _ _ _ _

4. Draw a pentagon.

Then draw a triangle <u>inside</u> the pentagon.

Draw a large hexagon.

Then draw a pentagon inside it.

Then draw a quadrilateral inside the pentagon.

Then draw a triangle inside the quadrilateral.

A.

D.   C.

E.

.B

1. Connect the points A, B, C, D, E <u>in that order.</u>
   Then connect E and A.

2. How many sides does this figure have? _ _ _ _ _

   What do we call this figure? _ _ _ _ _ _ _ _ _ _ _ _

3. Match each word with all the figures it names.

   quadrilateral

   pentagon

   hexagon

   circle

   triangle

Accuracy Quiz

1. Connect A and B.

A   B   P   Q   X   Y

2. Connect P and Q.

3. Draw line segment $\overline{XY}$.

4. Draw the line determined by A and Y.

5. What can we say about points A, B, P, Q, X, and Y? _ _ _ _ _
_ _

•O

U •   •V

M •   •N

•
W

6. Draw line segments $\overline{OW}$, $\overline{UN}$, and $\overline{VM}$.
   What happens to all three of these line segments? _ _ _ _ _ _ _ _
_ _

7. Draw all line segments determined by two of the points C, D, E, F, G.

C•

G•   •D

F•   •E

<u>Rays</u>

1. <u>Extend</u> each line segment in both <u>directions</u>.

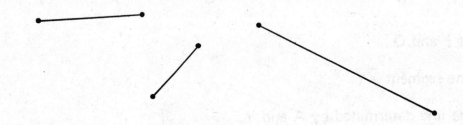

2. Extend line segment $\overline{AB}$ <u>beyond</u> endpoint B.

This figure, with only one endpoint "A" and extended in one direction, is a <u>ray</u>. We call it ray $\overrightarrow{AB}$.

3. Extend $\overline{PQ}$ beyond Q, and extend $\overline{PR}$ beyond R to form rays $\overrightarrow{PQ}$ and $\overrightarrow{PR}$.

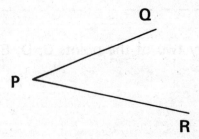

4. Draw a ray, and label its endpoint.

Angles

1. Extend $\overline{BA}$ beyond A to form ray $\overrightarrow{BA}$.

   Extend $\overline{BC}$ beyond C to form ray $\overrightarrow{BC}$.

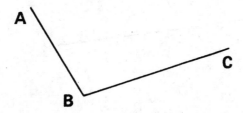

   This figure is an <u>angle.</u>

   We call it angle ABC.

2. Extend the <u>sides</u> of each angle.

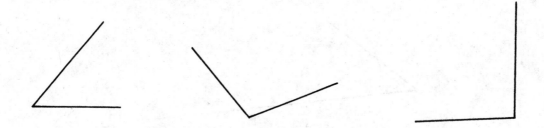

3. What is the point of intersection of the sides $\overrightarrow{XY}$ and $\overrightarrow{XZ}$ of the

   angle YXZ? _ _ _ _ _

   What is the vertex of the angle? _ _ _ _ _

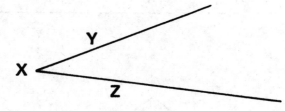

4. Name this angle. _ _ _ _ _

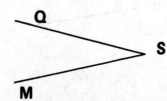

1. Name the segment in two different ways. _ _ _ _ _ _ _ _ _ _ _ _

2. Name the angle in two different ways. _ _ _ _ _ _ _ _ _ _ _ _ _ _

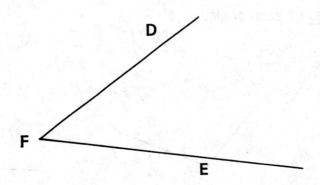

3. Name the triangle in six different ways. _ _ _ _ _ _ _ _ _ _ _ _

_ _

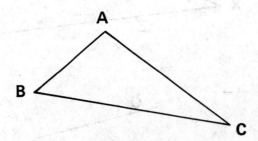

Congruent Figures

1. <u>Trace</u> one of the quadrilaterals.

Place the tracing over the other quadrilateral.

Can you make them match exactly? _ _ _ _ _

Two figures that match exactly are <u>congruent</u>. They have the same shape. They have the same size.

Are the quadrilaterals congruent? _ _ _ _ _

2. Trace one triangle and see if the triangles are congruent.

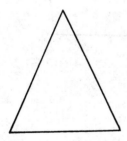

Are the triangles congruent? _ _ _ _ _ _

3. Are the line segments congruent? _ _ _ _ _

1. Extend the sides of each angle.

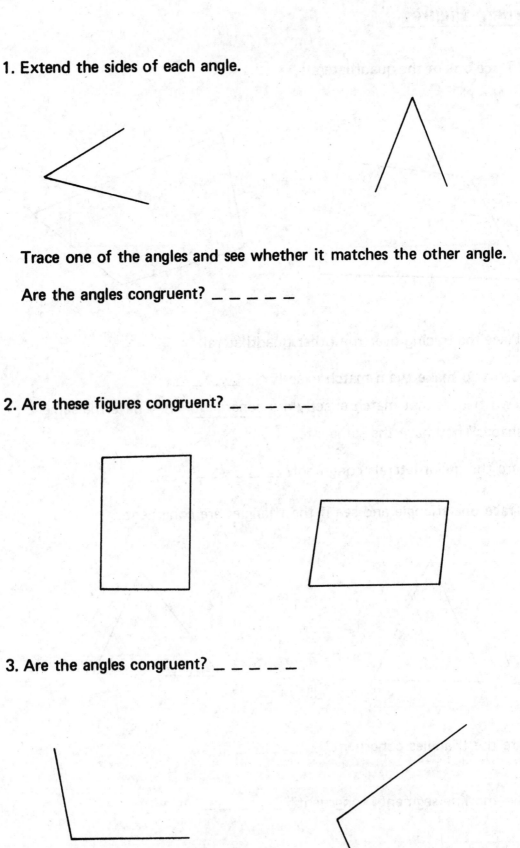

Trace one of the angles and see whether it matches the other angle.

Are the angles congruent? _ _ _ _ _

2. Are these figures congruent? _ _ _ _ _

3. Are the angles congruent? _ _ _ _ _

1. Trace one of these triangles.

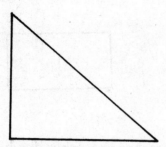 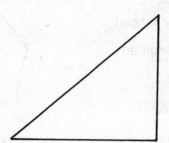

Place the tracing over the other triangle.

Can you match the triangles exactly? _ _ _ _ _ _

Turn your tracing over.

Now can you match the triangles? _ _ _ _ _ _

Are the triangles congruent? _ _ _ _ _ _

2. Are these figures congruent? _ _ _ _ _ _

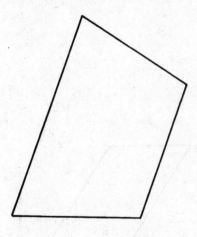 

3. Are these angles congruent? _ _ _ _ _ _

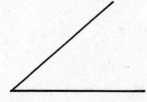 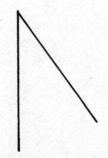

Practice Test

1. Match each word with the figure it names.

triangle

quadrilateral

line segment

angle

pentagon

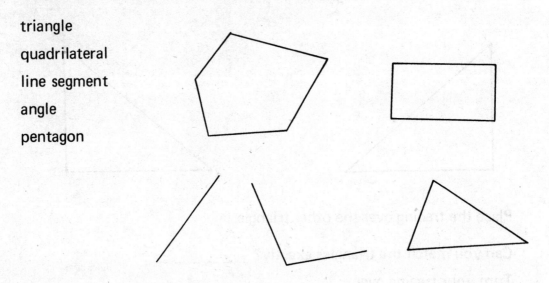

2. Draw the line determined by points A and B.

A.

•
B

3. Are the figures congruent? _ _ _ _ _

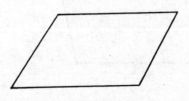   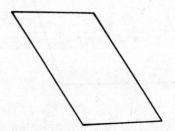

4. Draw $\overline{XY}$.

X.

•Y

5. Draw the triangle determined by A, B, and C.

A.

.C

B.

6. Given W, X, Y, and Z, draw all the line segments you can which connect two of them.

W.

.X

Y.

.Z

How many line segments did you draw? _ _ _ _ _ _

How many points of intersection are there? _ _ _ _ _ _

How many triangles do you see? _ _ _ _ _ _

7. Extend the sides of angle DEF.

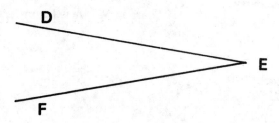

8. Draw a hexagon.

Key to Geometry® workbooks

The KeyTo TRACKER™

The Online Companion to the *Key to...*® Workbook Series

Save time, improve learning, and monitor student progress with The Key To Tracker, the online companion to the Key to... workbooks for fractions, decimals, percents, and algebra.

Learn more: www.keypress.com/keyto

Also available in the Key to...® series

Key to Fractions®

Key to Decimals®

Key to Percents®

Key to Algebra®

Key to Metric®

Key to Metric Measurement®

Key Curriculum Press

INNOVATORS IN MATHEMATICS EDUCATION

ISBN 978-0-913684-71-9